Basic Essentials®

Weather
Forecasting

Help Us Keep This Guide Up to Date

Every effort has been made by the editors to make this guide as accurate and useful as possible. However, many things can change after a guide is published—new products and information become available, techniques and theories evolve, etc.

We would love to hear from you concerning your experiences with this guide and how you feel it could be improved and kept up to date. While we may not be able to respond to all comments and suggestions, we'll take them to heart and we'll also make certain to share them with the editors. Please send your comments and suggestions to the following address:

The Globe Pequot Press
Reader Response/Editorial Department
P.O. Box 480
Guilford, CT 06437

Or you may e-mail us at:

editorial@GlobePequot.com

Thanks for your input, and happy travels!

A FALCON GUIDE®

BASIC ESSENTIALS® SERIES

BASIC ESSENTIALS®
Weather
Forecasting

Third Edition

MICHAEL HODGSON

ILLUSTRATIONS BY DEVIN WICK

FALCONGUIDE®

GUILFORD, CONNECTICUT
HELENA, MONTANA
AN IMPRINT OF THE GLOBE PEQUOT PRESS

FALCONGUIDE ®

Text and page design by Nancy Freeborn
Photo credits: figures 8, 9, 10, 11, 12, 13, 14, and 16 courtesy NOAA; figures 15 and 17 courtesy of the Plymouth State University Meteorology program; figures 24, 26, 27, 28, and 29 courtesy Wind & Weather

Library of Congress Cataloging-in-Publication Data is available.

ISBN-13: 978-0-7627-4223-3
ISBN-10: 0-7627-4223-2

Manufactured in the United States of America
Third Edition/First Printing

*To Therese, whose smile warms me
no matter what the weather*

Contents

Introduction ix

1 How Weather Happens 1

2 Understanding Clouds 13

3 Geographic Weather Variations 24

4 Forecasting Changes in Weather 34

5 Nature's Signs 43

6 Backyard Meteorology 49

Appendix 1: Glossary 55

Appendix 2: Recommended Reading and Web Sites 59

Appendix 3: Suppliers of Weather Instruments 63

Index 65

About the Author 69

Introduction

"Tonight's weather is dark, followed by widely scattered light in the morning. . . ."

—George Carlin, from his Hippi Dippi Weatherman routine

A number of years ago, actually more years than I care to remember, two friends and I were backpacking along the Appalachian Trail just north of Great Smoky Mountains National Park. We were on a weeklong backpack to celebrate the independence of our sixteen-year-old spirits.

Just after dawn on the second day of our adventure, the clouds, which had been stacking ominously for hours, released a deluge of water lasting for several hours. The deluge reduced to a steady drizzle that continued until evening. Soaked and somewhat dispirited, we trudged into camp, happening upon a crusty old traveler slouched against a pack that appeared as if it may have been a prototype for later cruiser-frame models.

I glanced pensively up at the lowering ceiling, which would soon turn to a dense mist, and then over to the old-timer who had either been ignoring our dripping entrance or was unaware of our presence—I sensed the former. Shuffling closer, not wanting to offend, I managed a somewhat unsure, "Mister, is this weather going to keep up, or do you think it'll be sunny tomorrow?"

With a cough the old-timer lifted the brim of his felt hat, gazed intently at me and my friends for what seemed an eternity, and then, without smiling, drawled, "Waell boys, I kin guarantee ya'll one thing fer sher. Come mornin' there's bound to be weather of some sort or the other. Wet or dry, ya'll don't got much choice in the matter, so why waste yer time frettin' over what ya cain't control." With those words of wisdom, the old-timer pulled himself upright, swung his pack onto one shoulder, tipped his hat respectfully in our direction, and then disappeared into the dusk and mist.

I'll never forget those words, and though they are for the most part filled with truth, there is one important element missing. While one cannot do much about the weather, by learning to read and understand changes in weather patterns and what those changes mean, one can experience the vast difference between blind reaction and reliable preparedness. Often, that difference alone may determine the margin of comfort and safety that separates disaster from adventure.

Continually practice keeping a weathered eye turned upward toward the sky. The more you are aware and the more you learn about what causes weather, the more perceptive will be your observations and the more accurate your guesses as to the weather's outcome. But never forget—predictions relative to weather are only educated guesses, never statements of fact. Always be prepared for the worst.

CHAPTER ONE

How Weather Happens

In the Northern Hemisphere warm (tropical) air moves north and cold (polar) air south, generally speaking. With that in mind one can expect warm fronts to be generated from the southern reaches and cold fronts from the northern. On weather charts and some maps, three types of air masses (figure 1) are often noted—maritime polar (mP), cold polar air that formed over the ocean; maritime tropical (mT), warm tropical air that formed over the ocean; continental polar (cP), cold polar air that formed over land.

With polar air masses the weather is apt to change abruptly, and as the air warms over land, it becomes turbulent, with associated cumulus clouds and often heavy precipitation. Tropical air is more stable since it is already quite warm, and while it often brings precipitation, the weather associated with a tropical air mass is apt to stay around for a while.

Both maritime and continental polar air influence local weather conditions around North America. In San Francisco, the cold maritime polar air that causes coastal fog during summer causes heavy rains in winter. Sometimes thundershowers form in the Sierra in summer, dropping precipitation on the western slopes. Continental polar air causes turbulent weather conditions and rains in the Great Lakes region during summer and heavy snows in the southeastern reaches of the lakes in winter.

Maritime tropical air brings with it humidity and extreme heat to the East during summer months. In winter it brings heavy rains.

As an air mass moves around the earth's surface, contact with other air masses is inevitable. These points of contact are called fronts, and understand-

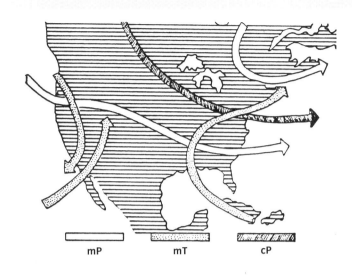

Figure 1. Movement of air masses: maritime polar (mP), maritime tropical (mT), and continental polar (cP).

ing how fronts interact is critically important to understanding how weather happens.

Fronts don't just suddenly appear out of nowhere. Forces, called pressure systems, are at work, pushing and pulling the various cold and warm air masses at will. High-pressure cells of the Northern Hemisphere create winds that rotate in a clockwise direction, whereas winds associated with low-pressure cells rotate counterclockwise. The reverse is true south of the equator.

High-pressure systems are associated with relatively cold air, whereas low-pressure systems are associated with warm air. Since cold air is heavier than warm air, it exerts a downward pressure on the earth. This pressure is registered by a rise in barometric pressure on a barometer. Conversely, warm air rises, resulting in a releasing of pressure on the earth's surface. This pressure is registered by a lowering in barometric pressure on a barometer.

Cold and warm fronts never mix. They displace each other, forming associated but separate independent systems that, in actuality, alternate with each other. The weather observer can determine changes in weather patterns by noting changes in barometric pressure, coupled with prevailing winds. There is more on how to observe weather trends in chapter 4.

Although wet, turbulent weather tends to be associated with low-pressure systems, or depressions, and fair weather with high-pressure systems, or ridges, exceptions to the rule do happen. The weather associated with either a cold or warm front is just as likely to be fair or poor when either is in firm control, depending on the direction of the winds brought by the prevailing system. If the winds have passed over a significant body of water, either system—high or low—can bring precipitation.

How quickly barometric pressure changes as a particular system moves in is also an indication as to how long in duration and how severe any accompanying storm may be. A rapid drop in barometric pressure indicates that any storm brought in by the low-pressure system will likely be short. If the drop in pressure is slow and steady, expect the accompanying storm to be long and severe.

Rising pressure usually brings a fair change to the weather. If, however, the pressure was very low to begin with, it is possible for a period of intense rain squalls to move through before any sunshine dances among the leaves. A rapid rise in pressure will often bring high winds due to unstable atmospheric conditions.

FRONTS

When air masses of different temperatures and densities meet each other, they do not mix well. Think of a bunch of heavy-metal rockers crashing a Dolly Parton benefit—there's bound to be tension. When this happens, battle lines are drawn, and a front is defined. There are three types of fronts: warm, cold, and occluded.

Warm Fronts

Warm air is far more stable than cold air. It is also more moist, with lower ceilings and poorer visibility, even if there is no appreciable precipitation. While weather associated with warm fronts is typically less severe than that attributed to cold fronts, the weather is frequently longer lived—rain may last several days or more.

Warm fronts move at a relatively slow speed, 10 to 20 miles per hour, and may be anticipated as much as two days in advance by a consistent sequence of cloud formations and a drop in barometric pressure (altimeter needle gains elevation). The usual cloud formation announcing an incoming warm front (see figure 2) is preceded by cirrus clouds followed by, in succession, cirrostratus,

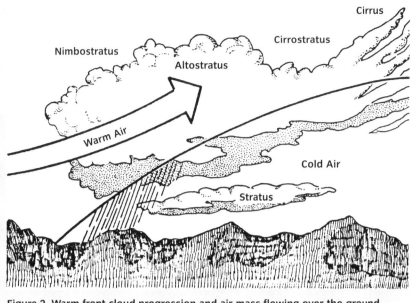

Figure 2. Warm front cloud progression and air mass flowing over the ground.

altostratus, and finally nimbostratus. (See chapter 2 for specifics on clouds.)

As a warm front moves in, it slowly displaces the cold air that came before by rising above it and slowly warming it. The gentle slope of the warm front as it is forced up and over the cold air, gradually cooling to dew point, results in the predictable formation of clouds as described in the preceding paragraph.

Cold Fronts

Cold air is more unstable than warm air and consequently very active. High ceilings and good visibility are associated with cold fronts, unless there is precipitation. Weather associated with a cold front is often severe and violent in nature. Typically, though, weather conditions associated with cold fronts are also shorter in nature than those associated with warm fronts. Cold fronts move at a speed of approximately 25 to 35 miles per hour and generally originate in the north or west when forming in the Northern Hemisphere.

As a cold front moves in, it pushes under the warm air, which rises and cools. If dew point is reached by this rising and cooling, precipitation occurs, often quite heavily. Cold fronts frequently arrive with very little warning. Nim-

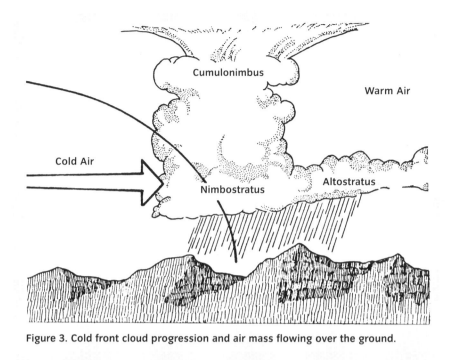

Figure 3. Cold front cloud progression and air mass flowing over the ground.

bostratus and cumulostratus are both rain clouds that may be generated by a cold front. Both are usually preceded by altostratus, possibly after an advance squall line of thundershowers has passed through. See figure 3.

Occluded Front

An occluded front (figure 4) occurs when one air mass gets caught between two other air masses and is forced off the ground. What actually happens depends almost entirely on temperature. As a warm front moves through, with a cold air mass in front and a cold front behind chasing it, several things can happen. If the air behind the pursuing cold front is colder than the air in front of the warm air mass, then the advancing cold front will actually lift both the warm and cold air in front of it. Weather associated with this type of front is usually squalls with thunder and lightning.

If the temperatures are reversed, with the colder air in front of the warm air, the pursuing cold front will be forced up over the warm front. The weather is likely to include heavy precipitation.

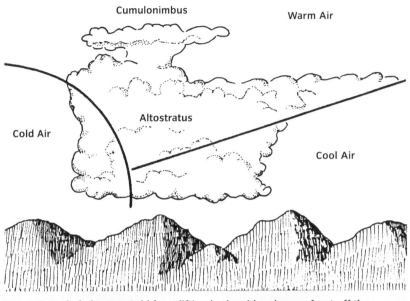

Cumulonimbus

Warm Air

Cold Air

Altostratus

Cool Air

Figure 4. Occluded Front: Cold front lifting both cold and warm front off the ground.

THUNDERSTORMS AND LIGHTNING

Extreme and violent vertical movements of air, know as updrafts, often lead to thunderstorms. This uplifting usually creates dramatic cumulus and cumulonimbus cloud formations that rise vertically, sometimes as high as 70,000 to 75,000 feet. The uplift of air can be caused by heating of the air from the ground (common in the Midwest), weather associated with a cold front, or temperature differences caused by the meeting of land and sea (common in the Gulf and southern Atlantic states).

Because temperature differences are the most common cause of thunderstorms, thunderstorms usually occur in the afternoon, when temperature differences between land and air or sea are the most extreme. Since temperature differences between land and sea are not as dramatic in the North Atlantic or West Coast (Pacific Ocean) regions, thunderstorms in these locations are not as frequent.

A thunderstorm forms as rising warm air cools and condenses. The higher the cloud stacks, the greater the level of cooling, and soon precipitation begins. As the falling rain and ice crystals cool the air within the cloud, the temperature difference between the cloud and the surrounding air begins to equalize, leading to the creation of downdrafts and heavy precipitation. The rush of cold air fans out ahead of the storm, sometimes as much as 2 to 3 miles, and is a good indicator of an impending thunderstorm.

As the downdrafts increase in intensity, the cloud becomes nothing more than downward moving air. Cooling of air ceases, and since no air is rising and condensing, the rain tapers off.

Lightning is caused by the attraction of unlike electrical charges within the storm cloud or the earth's surface. The friction caused by rapidly moving air particles, churning from the violent updrafts and downdrafts, leads to a building up of strong electrical charges. As the electrical pressure builds, charges between parts of the cloud or from the cloud to the earth are released, taking the form of lightning.

Up to thirty million volts can be discharged by one single lightning bolt, and it is this power, or explosive heating of the air, that causes the compressions of thunder.

Figure 5. Place yourself in the middle of a tree cluster during a thunderstorm.

It is possible to roughly judge the distance of an approaching storm by observing the lightning's flash followed by the resounding boom of thunder. For this method count slowly, "one-one thousand, two-one thousand, three-one thousand," and so forth. This will approximate one second elapsed for each thousand counted. Once the lightning has flashed, begin to count. When you hear the thunder, stop counting. Every five seconds of elapsed time indicates 1 mile of distance. In other words, if the count reaches, "seven-one thousand," it is safe to assume that the approaching thunderstorm is approximately 1½ miles distant.

Thunderstorms are dangerous companions when you are traveling on a mountain peak. If a thunderstorm approaches when you and your hiking party are on an exposed peak or ridge, take the following precautions:

- Get off the ridge or peak if at all possible. Even reaching a few feet below the highest point around you is better than not moving at all.

- Get away from your pack, as the metal in the pack will conduct electricity.

- Position yourself on a dry surface, preferably insulated, such as a sleeping pad. If the sleeping pad has become soaked in the rain, however, it will be useless, as water conducts electricity.

- Do not lie down or sit. Rather crouch on the balls of your feet with your feet close together. The idea is to minimize the available surface area through which possible ground currents from nearby lightning strikes may move.

- Do not huddle next to a single tree. That is rather like hugging a lightning rod and expecting to be safe. Choose a cluster of trees instead and place yourself in the middle (figure 5), preferably in an open area.

- Spread the group out, at least 25 to 30 feet apart. If lightning does strike, the idea is to minimize the potential damage and injury. With the group spread out, chances are only one person will be injured, if, at all, and this leaves the rest of the party to provide lifesaving assistance once the storm moves on.

- Avoid depressions or caves. Such areas usually have moisture present, which makes them more susceptible to conducting electricity.

- If you are caught on a lake in a thunderstorm, assume a crouching position toward the middle of the boat. Try to minimize your contact with wet objects.

HURRICANES

Hurricanes are tropical cyclones, low-pressure systems. Fortunately, the typical outdoorsperson has little to worry about from hurricanes due to their rarity of occurrence in recreational areas away from the ocean. Hurricanes form only over open ocean areas that are covered by an extremely warm and moist air mass—which is why they are usually called tropical cyclones. Hurricanes will always break up and lose intensity within hours of moving over a large land mass. Once inland, the hurricane's major effect is near-torrential downpours, but not, thankfully, the severe winds felt in coastal regions. It is virtually impossible for the amateur weather observer to anticipate the approach of a hurricane, other than by listening to the National Weather Service. The associated clouds and lowering of atmospheric pressure that precede the arrival of a hurricane are decidedly similar to those of a warm front.

TORNADOES

Tornadoes are perhaps the most violent and intense of all known storms, with winds in excess of 300 miles per hour recorded in the vortex. Man-made structures seem to explode as the extremely low pressure within the vortex of a tornado causes the normal pressure trapped within structures to expand rapidly, ripping buildings to shreds from the inside out.

Violent updrafts, recorded between 100 to 200 miles per hour, within the center of the funnel cloud have been known to suck anything within their path hundreds of feet in the air before hurling the swirled objects some distance away. Over bodies of water, funnel clouds lift water into the air, creating what is known as a waterspout.

Most tornado-related injuries and deaths are caused by debris flying through the air and falling objects. If a tornado approaches, take the following precautions:

- If outside, curl up in a tight ball in a drainage ditch, a culvert, or any other depression that will protect you. Cover your head. Be very alert to the potential for flooding.

- Do not attempt to outdrive a tornado! Get out of your vehicle and curl up in a tight ball in a ditch or depression away from the vehicle.

- If indoors, leave the windows closed and seek shelter in the basement, if there is one, or in a room located in the center of the home with no windows or the fewest number of windows possible. Curl up against a wall and cover your head.

- If in a mobile home, an RV, a barn, or a small building that you know is not firmly anchored to the ground, get out. The tornado will toss it about like a house of straw, and you are in more danger being inside than you are outside.

RAIN AND SNOW

Not every cloud produces precipitation, and even in those clouds that do, the conditions must be just right for rain or snow to fall. Every cloud is made up of millions of tiny droplets of moisture or ice crystals, so small that each droplet or crystal is easily held aloft by moving air even though the pull of gravity on them is constant. The only way each droplet or crystal can fall to earth is if it gains enough size so that the pull of gravity overcomes the lift of moving air. This process of growth is referred to as coalescence.

Coalescence can occur in a variety of ways. The two main methods of formation are described as follows:

1. Droplets of rain collide, becoming bigger. As they become bigger, they are not whipped around quite so much by the movement of air, and because of a larger surface area, they collide with more and more droplets until they become large enough to fall as rain.
2. When water droplets and ice crystals occur in the same cloud, as they do in cumulonimbus, some of the water will evaporate and then condense again on the ice crystals. As the crystals grow, they fall toward earth as snow or ice pellets that melt in the warmer air of lower elevations, finally splashing down as raindrops.

Snow forms only when supersaturated air in the cloud is cold enough (at or below 10 degrees F) and when there are sufficient small particles of debris in the cloud upon which water vapor can crystallize. If supersaturated air within the cloud is colder than –38 degrees F, snow can form without the presence of debris.

DEW AND FROST

Dew is not precipitation. It is water vapor that condenses on the surface of solid objects (such as a tent, grass, or a pack) that have cooled below the condensation point of the air that is in contact with them. Dew is usually heavier near bodies of water, in valley bottoms near streams or rivers, and on clear nights when cooling by radiation is at its greatest.

Frost is a type of dew, but it forms only when surface temperatures are at or below freezing. Frost forms as water vapor crystallizes instead of condensing into water droplets.

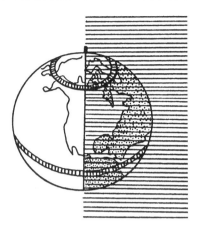

Spring equinox: North and South Poles are equidistant from the sun.

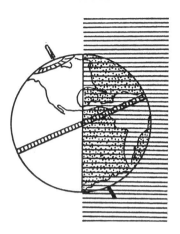

Summer solstice: In the Northern Hemisphere, the North Pole is angled toward the sun and days are longer, hence warmer.

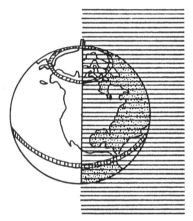

Fall equinox: North and South Poles are equidistant from the sun.

Winter solstice: In the Northern Hemisphere, the North Pole is angled away from the sun and days are shorter, hence colder.

Figure 6.

THE SEASONS

The seasons are created by the earth's orbit around the sun, coupled with the fact that the earth's rotation on its axis is angled at 23.4 degrees to the plane of its orbit. Say what? It's not as complicated as it sounds. First, think of a globe: The earth is constantly spinning, one rotation every twenty-four hours, on an axis. (Note that every accurate globe is angled at 23.4 degrees.) The earth also orbits the sun, one rotation every 365 days. The orbit is maintained on a constant plane, which means that the earth's relative position to the sun takes on three separate angles to the sun as it maintains its orbit.

In the Northern Hemisphere summer, the North Pole is angled toward the sun. More sun is hitting the northern reaches of the globe, meaning that days are longer and the sun's rays are striking the earth more directly in the north. This is why summer is warmer than winter. In winter the South Pole is angled toward the sun, which reduces the sun's contact with the earth in the Northern Hemisphere and also diffuses the sun's warming rays, effectively lowering temperatures. In spring and fall the North and South Poles are relatively equidistant from the sun. The seasons are reversed in the Southern Hemisphere. See figure 6.

CHAPTER TWO

Understanding Clouds

WATER CYCLE

Although not specifically an integral part of weather forecasting, the water, or hydrological, cycle plays a vital role in the formation of clouds and the eventual precipitation that returns the moisture to earth (figure 7). Moisture in the air comes from many sources. Water evaporates into the atmosphere from oceans, seas, lakes, rivers, ponds, wetlands—any open body of water that comes into contact with air. As this warm, moisture-laden air cools, either adia-batically (due to a risse in elevation) or by coming into contact with cooler air, the water vapor in the warm air condenses, forming clouds, and then inevitably falls back to earth as precipitation—rain, snow, sleet, hail, etc.

CLOUD FORMATION

In general, cloud formations are made up of water droplets and ice crystals that are held aloft by turbulent air movements. As the droplets of moisture or crystals of ice within clouds increase both in number and size, they become too heavy for the air to support and consequently fall to the earth as precipita-tion. These water droplets and ice crystals all got where they are the same way, by the cooling of air below its saturation point, or the point at which water vapor condenses.

Clouds can be formed any number of ways, but the most common are as follows:

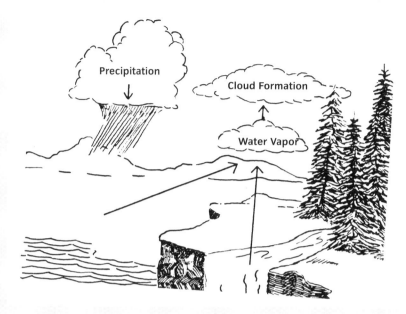

Figure 7. Water, or hydrological, cycle: Water evaporates, rises on warmer air, cools, condenses, forms clouds, and then falls back to earth as precipitation.

- As the earth radiates heat during a clear night, fog may form near the earth's surface due to warm air contact with a cooling earth.

- Coastal areas are very familiar with the fog and mist caused by warm ocean air moving over a cooler land surface and condensing. The same conditions can be associated with any large body of water.

- As a cold front pushes under a warm front, clouds form as the warm air, now cooling as it rises, condenses.

- Precipitation falling from high clouds sometimes cools the air below, causing lower elevation clouds to form.

- Clouds form as air cools and warms adiabatically from being pushed over a mountain ridge or peak.

WHAT DO CLOUDS MEAN?

Through the consistent observation of clouds, the weather watcher can draw some relatively accurate conclusions regarding shifting weather patterns. If clouds are massing and generally increasing in size and density, the weather is possibly changing for the worse. As clouds begin to move faster across the sky, this pattern indicates changes in wind velocity and in barometric pressure—your clues that a storm approaches.

Since varying cloud formations take on different appearances and altitudes, it is possible, although sometimes difficult, to observe the movements of two or more different cloud formations across the sky. Cloud formations that are moving in separate directions and at different altitudes often announce an impending storm.

Small, black cumulus clouds, called scud clouds, often bring rain. Scud clouds that are seen sweeping beneath a dark or darkening stratus layer usually indicate an impending storm with accompanying winds.

Finally, observe cloud movements and wind direction together. One may assume that if both the cloud movement and wind direction are the same, any weather will come from the direction the wind is blowing.

TYPES OF CLOUD FORMATIONS

Cirrus Clouds

Made up predominantly of ice crystals, cirrus clouds (figure 8) often are associated with other cloud formations, especially cirrocumulus.

Cirrus clouds are arguably some of the most beautiful of the cloud formations, leaving milky white swirls and curls etched across the sky. Mare's tails are a type of cirrus cloud that serve as a warning of an approaching warm front and a possible storm.

Cumulus Clouds

Cumulus: Often referred to as heap clouds, cumulus clouds are typified by heaped or fluffy formations.

Cirrocumulus: High-level (usually around 25,000 feet) heap clouds very often seen combined with cirrus clouds. Cirrocumulus clouds (figure 9) indicate a condition of unstable air and may lead to precipitation before long.

Altocumulus: Medium-level (usually hovering around 8,000 feet) fleecy or puffy clouds, similar to dense cirrostratus, but without any telltale halo. When viewed in the early morning, altocumulus clouds usually indicate thunderstorms or precipitation within twenty-four hours—often that afternoon. See figure 10.

Cumulonimbus: Often-massive cumulus with a broad base ranging from 3,000 feet upward to 16,000 feet; even 65,000 feet is not unusual. Top is fuzzy or

Figure 8.
Cirrus clouds.

Figure 9.
Cirrocumulus clouds.

Figure 10.
Altocumulus clouds.

Figure 11.
Cumulonimbus clouds.

Figure 12.
Fair-weather cumulus
clouds.

Figure 13.
Cumulus congestus
cloud.

anvil shaped. Heavy downpours, coupled with hail, lightning, and thunder, are standard fare. See figure 11.

Fair-weather cumulus: Low-level (up to 4,000 feet) heap clouds that often form in the late morning or early afternoon. Clouds are not very dense, are white in color, and are well separated from one another. These clouds form when the air mass is stable and being warmed by the earth's surface. See figure 12.

Swelling cumulus: Medium-level (around 15,000 to 20,000 feet) heap clouds that also form in the late morning or early afternoon. Weather observers often refer to these clouds as giant heads of cauliflower because their bottoms are flat and their tops are bumpy. Swelling cumulus indicates unstable air masses. These clouds are quite common in the deserts of the Southwest.

Cumulus congestus: High-level (stacked from 5,000 up to 30,000 feet) heap cloud formed by massive uplifting of heated air within an unstable air mass. Its top is still bumpy and forming, marking the major difference between this cloud and the precipitating cumulonimbus. See figure 13.

Stratus Clouds

Stratus: Stratus means layered, essentially formless, with no real defining base or top. Fog is a type of stratus cloud that lies close to the ground and is caused when the earth's surface cools. This cooling effectively lowers the air temperature, resulting in condensation.

Cirrostratus: High-level (usually around 20,000 feet) veil-like cloud formations

**Figure 14.
Cirrostratus and cirrocumulus clouds.**

Figure 15.
Altostratus clouds.

Figure 16.
Nimbostratus clouds.

Figure 17.
Lenticular cloud.

composed of ice crystals and often spreading out over a very large surface area. Halos are very often observed in cirrostratus clouds, indicating a lowering of the cloud ceiling and possible precipitation within forty-eight hours. See figure 14.

Altostratus: Medium-level clouds (usually hovering around 8,000 feet) that are flat or striated, dark gray in color. A darkening of the cloud cover indicates possible precipitation within forty-eight hours. See figure 15.

Nimbostratus: Low-level, dark and thick clouds, often without any real defining shape. Their ragged edges, known as scud, produce steady precipitation. See figure 16.

Lenticular: Well known in the Sierra Nevada, lenticular clouds (figure 17), formally labeled *altocumulus lenticularis,* are lens-shaped clouds that form at or around mountain peaks due to a cresting, or a wave in the airstream passing over the peak. As air is forced up the mounting ridge, it cools, condenses, and forms into a cloud near the peak. As the air passes over the peak and heads back down the ridge, it warms, and moisture evaporates. Since condensation is only occurring at the peak, the cloud forms only at the peak, and even though winds are whipping through it, the cloud remains stationary.

ANTICIPATING WEATHER FROM CLOUD FORMATIONS

The following are some generally accepted guidelines to help anticipate weather changes by observing cloud movements and sequences.

Cold Front

The rapid formation of altocumulus followed in quick succession by cumulostratus and finally cumulonimbus usually indicates an incoming cold front.

Warm Front

Cloud cover that darkens, beginning with cirrostratus, then altostratus, and finally nimbostratus, over a twenty-four-hour period indicates an approaching warm front and impending rain. Quite commonly, the longer the sequence of cloud formation and darkening, the longer the storm.

Precipitation

If cirrostratus, altostratus, and stratocumulus are all visible, the probability of precipitation within twenty-four hours is high. If the clouds are whipping about in different directions, the precipitation is likely to be heavy and long in duration.

IDENTIFYING CLOUDS

Still having trouble identifying clouds? Identifying clouds takes practice, practice, and more practice. At first, just try systematically working your way through identification of each kind. Soon cloud identification will become second nature.

- First determine if the clouds are heap, layered, or precipitating. (Hint: If you are getting wet, the cloud is precipitating.)

- If they are heap clouds only, then estimate the height of the base and top. If the top is no higher than 10,000 feet, then it is fair-weather cumulus. If the top is no higher than 25,000 feet, then it is swelling cumulus. If the top is above 25,000 feet, then it is cumulus congestus.

- If the clouds are layered only, then estimate the height of the base and top. If the cloud is formed between 1,000 to 10,000 feet, it is stratus. If the cloud is formed between 10,000 to 20,000 feet, it is altostratus. If the cloud is formed above 25,000 feet, it is cirrostratus.

- If the clouds are both layered and heaped in shape, then estimate the height of the base and top. If the clouds are formed between 1,000 to 10,000 feet, they are stratocumulus. If the clouds are formed between 10,000 to 20,000 feet, they are altocumulus. If the clouds are formed above 25,000 feet, they are cirrocumulus.

- If the clouds are precipitating (rain or snow), then study the level of precipitation. It does not have to reach the ground to be considered precipitation. Remember cirrus clouds: high-level clouds above 25,000 feet that have an apparent white trail, much like a veil, dropping from the sky, which disappears as it evaporates. If the showers are steady, fluctuating between light and medium intensity, or just heavy mist, then the cloud is nimbostratus. If the showers are on again, off again with heavy- to medium-level precipitation, then the cloud is cumulonimbus.

A QUICK GUIDE TO FORECASTING FROM CLOUDS

By combining information regarding temperature changes, barometric pressure changes, and changes in wind direction, it is possible to read clouds and gain a general picture of what may be happening with the weather. Just remember, nothing in weather is ever guaranteed. In general, it is fair to assume the following:

The weather will get worse if:

- high and isolated cloud formations begin to get thicker, increase in number, and lower in elevation.
- clouds that are moving quickly across the sky begin to get thicker and lower in elevation.
- clouds are moving in all directions.
- clouds begin to darken noticeably on a western horizon.
- the piles of clouds you've been watching start to stack in huge, vertical columns before noon in summer.
- clouds that are puffy begin to form into sheets and lower in elevation.

The weather will get better or remain fair if:

- the morning fog burns off before noon.
- the cloud cover begins to break up noticeably.
- the cloud ceiling rises in elevation.
- sheetlike clouds begin to bunch up, and breaks of blue sky begin to show through.

What can you tell from particular cloud formations and their changing patterns? Plenty if you are observant, which means watching the cloud formations over a period of a few hours or more. I keep telling you to keep a weather eye skyward, don't I? Just don't trip over your boots!

You may assume the following generalizations about clouds and their meaning regarding changing weather patterns:

- If cirrus, cirrostratus, or cirrocumulus clouds stay thin, wispy, and white, the weather will likely remain fair for twenty-four hours.
- If cirrus, cirrostratus, or cirrocumulus clouds start to thicken, join together into sheets, and lower, then precipitation can be expected within the next twenty-four hours.
- If altocumulus clouds begin to form sheets and/or if a corona (a luminous circle or ring formed around the sun or moon) begins to shrink, rain or snow can be expected within twelve hours, with warmer, drier weather to follow.
- If altocumulus clouds scoot across the sky from the west or northwest, you can anticipate an impending hard rainstorm or storms that are short in duration, to be followed by clear, cooler weather.

- If altocumulus clouds begin to form in the mid-morning hours on a humid summer day with a series of turrets rising up from one cloud, you should expect thunderstorms and gusty winds by afternoon.

- If altocumulus clouds remain patchy with clearly observed blue sky, and if they are moving in the same direction as the surface winds, you can anticipate fair weather for the next twelve to twenty-four hours.

- If altostratus cloud layers begin to darken and lower, plan on weather with steady precipitation and poor visibility over the next twenty-four hours.

- If stratocumulus clouds remain scattered and drift slowly with the prevailing wind, the weather will likely remain unchanged for at least twenty-four hours.

- If stratocumulus clouds become gray and begin to form into overcast layers, expect rain showers on and off.

- If stratocumulus clouds form in a bank to the west or the northwest and winds are coming from the south or southwest, anticipate a short but fierce storm, followed by clearing and colder weather.

- If nimbostratus clouds become dark and begin developing from lowering altostratus formations with winds coming from the southeast, rain is just around the corner or up the valley. Likely, though, the precipitation will be short lived, and warmer weather with clearing skies will follow.

- If cumulus clouds do not develop until afternoon or float alone in the sky without stacking, fair weather for the next twenty-four hours is likely.

- If cumulus clouds develop in the morning and begin stacking throughout the afternoon, especially on warm, humid days, scattered showers by late afternoon are likely.

- If cumulonimbus clouds form isolated masses moving from the south or the southwest, scattered thundershowers can be expected, most likely with lightning, possibly hail, and even high winds when the storm passes through.

- If cumulonimbus or very heavy cumulus clouds form solid banks along the western horizon, northwest, or north and winds are out of the south or southwest, anticipate a heavy thundershower or two to pass through before the weather clears.

- If cumulonimbus clouds turn black with huge, rounded bases that resemble lightbulbs in shape and look as though they are continuing to lower in elevation, you should watch for severe thunderstorms and even tornadoes.

CHAPTER THREE

Geographic Weather Variations

Although the weather may be warm and sunny on a ridge, temperatures can be downright cold in the valley below. How is this possible? Attribute temperature and humidity fluctuations to microclimate influences. The differences in temperatures and conditions in various microclimates are created by the different degrees of solar warming, radiation, and evaporative cooling. Knowing what these are and how they work can make the difference between choosing to spend a chilly night huddled in your sleeping bag within a cold sink (a cool air pool created as warm air rises from earth and is displaced by denser, cooler air at night) or basking comfortably several hundred feet away under moonbeams. Of course, if the weather is foul, microclimatology is a moot point—foul weather is a constant with no microclimate sanctuaries.

ADIABATIC FLUCTUATIONS IN TEMPERATURE

Air cools approximately 5.5 degrees F for each 1,000 feet of elevation gain (figure 18). It warms at approximately the same rate during elevation loss. As air meets a mountain and is forced over it, the air pressure reduces, and the temperature lowers as the air expands. Once the air pours over the peak and begins to descend, the temperature rises as the air pressure increases, and the air mass gets compressed. This adiabatic warming or cooling doesn't happen only on mountain peaks. Air rising and lowering through the atmosphere is also subject to adiabatic warming and cooling.

Condensation throws a monkey wrench in the process. Just as adiabatic cooling lowers the temperature of air, any condensation that may occur warms

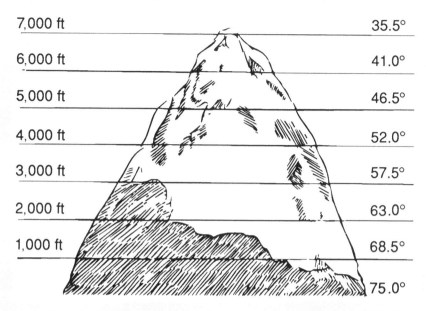

7,000 ft	35.5°
6,000 ft	41.0°
5,000 ft	46.5°
4,000 ft	52.0°
3,000 ft	57.5°
2,000 ft	63.0°
1,000 ft	68.5°
	75.0°

Figure 18. Air temperature rises and cools at an even 5.5 degrees per 1,000 feet when no moisture is present.

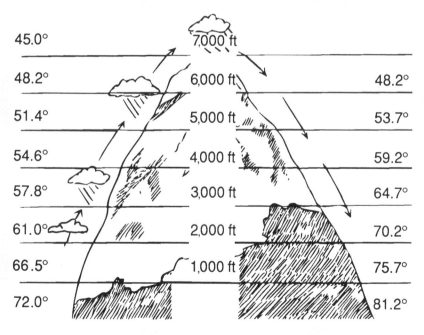

45.0°	7,000 ft	
48.2°	6,000 ft	48.2°
51.4°	5,000 ft	53.7°
54.6°	4,000 ft	59.2°
57.8°	3,000 ft	64.7°
61.0°	2,000 ft	70.2°
66.5°	1,000 ft	75.7°
72.0°		81.2°

Figure 19. Add condensation/moisture to the mix and the air cools at a slower rate (approximately 3.2 degrees per 1,000 feet) and then warms at a slower rate, too, until the moisture evaporates, illustrating adiabatic cooling and warming.

it. The net effect on humid air being pushed up over a mountain peak is an average cooling of the temperature at 3.2 degrees F per 1,000 feet. Once at the peak, the air begins to warm as it descends, and no further condensation occurs. Adiabatic warming of this air happens at the normal 5.5 degrees F for each 1,000 feet of elevation loss. See figure 19.

What does all this mumbo jumbo mean? Let's look at a 7,000-foot mountain as an example. As moisture- laden air climbs the mountain, it cools and condenses, often creating a localized weather system of rain showers and mild winds. For the sake of the illustration, we will assume that the condensation and precipitation begins at 2,000 feet. If the temperature at ground level was 72 degrees F, then at 2,000 feet the temperature has been lowered to approximately 61 degrees F (5.5 degrees F for every 1,000 feet). Between here and the peak, the air begins to condense, thus cooling at a slower rate (3.2 degrees F per 1,000 feet), resulting in a peak temperature of 45 degrees F. As the air mass slides over the peak and begins to descend, the condensation evaporates quickly, within 1,000 feet. At 6,000 feet the temperature has risen 3.2 degrees (the first 1,000 feet were warming at a moist adiabatic rate) to 48.2 degrees. From 6,000 feet down to ground level, however, the air will warm at a dry adiabatic rate since the relative humidity is decreasing, resulting in a final ground temperature of 81.2 degrees F—warmer than when the air mass was forced over the mountain beginning at the same elevation on the other side.

CHINOOK WINDS

As a boy living in Calgary, Alberta, located at the base of the Canadian Rockies, I remember chinook winds well. This weather phenomenon could raise the cold winter temperatures of the morning into a balmy but very dry afternoon, causing snow to vanish almost magically—confusing indeed to a young boy expecting to go sledding, but ideal sandbox weather nevertheless. Chinook winds occur in the Sierra, the Rockies, and many other mountain locations around North America and the world. Some even carry their own label, such as the famous Santa Ana winds of southern California.

These winds and the accompanying fluctuations in temperature are caused by strong low- or high-pressure systems that force an air mass over a mountain range. As the air mass climbs the mountain, it cools adiabatically as described above, often dropping snow or rain on the windward side of the mountain. Once over the mountain peaks, its moisture depleted, the air mass streams back down the mountain. The warming effect caused by the condensation, coupled with adiabatic warming through elevation loss, sends hot, dry air roaring through the lowlands (temperature climbs of as much as 40 to 45 degrees F within fifteen- to thirty-minute spans have been recorded in Alberta). The winds that blow are fairly constant, usually not over 20 to 25 miles per hour (although winds have been recorded at near the 100 mph mark), and they blow twenty-four hours a day since they are governed by the driving force of a pressure system, not by thermal heating and cooling.

MOUNTAINS AND VALLEYS

Hike anywhere in any mountain range and you will note one fairly constant feature: Wind blows up the mountain during the day and down the mountain at night (figure 20). Why? Two forces are at work, both caused by radiation as the earth's surface loses more heat than it absorbs during a clear night. As the ground gives off heat, the heat rises and the earth cools. The air close to the ground becomes colder more quickly and, since it is denser than warm air, begins to flow, rather like water, downhill—toward valley bottoms and desert floors. As the cool air rushes down the mountain (this is the breeze you feel), it displaces the warm air, forcing it upward, keeping a cycle of air going. As the warm air gets forced out, the cool air pools and collects in the lower reaches, creating places often referred to as cold sinks.

In broad valleys covered with dense meadow grasses, frost pockets may form, even though the temperature just 500 to 1,000 feet above is a comfortable 50 degrees. The broader the valley and the more vertical the surrounding mountainous ridges, the more marked the temperature contrasts will be. The narrower the valley, the less dramatic the temperature differences between top and bottom will be since the walls tend to radiate heat from one to another, effectively trapping potential heat loss from the valley bottom.

As the early morning sun begins to warm the valley, the winds reverse themselves and begin rushing up the mountain as the warm air begins to rise. Remember, however, that the air can only be warmed if the sun hits the ground below it, radiating heat upward. This explains why the side of the valley that is getting the sun first will experience upslope winds, while the side of the valley still in shade continues to experience cool downslope winds. See figure 21 for an illustration of valley air circulation during the day and night.

Cloudy conditions or a presence of strong winds in the area will alter the topographical variations in wind and temperature patterns. Cloudy conditions act as an insulator, reducing radiant heat loss and holding a thermal layer of air closer to the earth's surface. Windy conditions may overpower the gentle

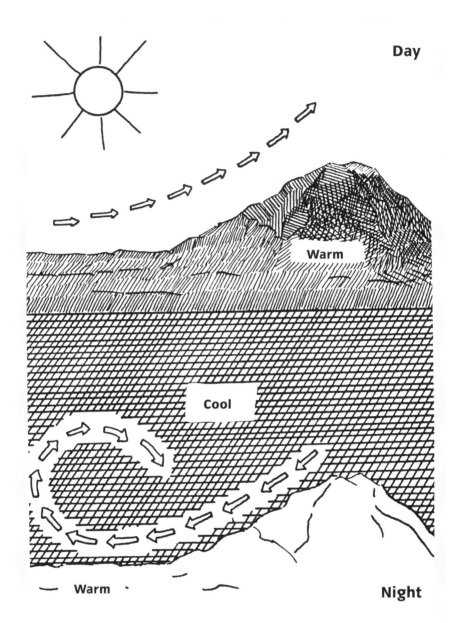

Day

Warm

Cool

Warm

Night

Figure 20. Air flows up the mountain during the day and down the mountain at night.

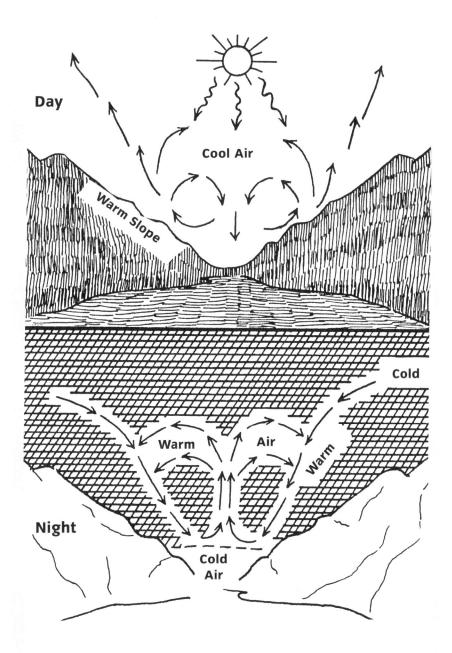

Figure 21. Valley air circulation during the day and night.

nature of upslope or downslope winds and also alter the potential temperature variances.

Winds that flow up and down a valley or canyon floor operate in a similar manner to winds that flow up and down a valley's walls. As temperatures fluctuate between higher elevations (mountains, ridges, or foothills) at one end of the canyon or valley and the lower elevations (desert, plains), winds also fluctuate in intensity and direction. Canoeists and rafters know this well, as many an afternoon wind blowing upstream from the heated elevations below has thwarted efforts to paddle successfully into it. Keep in mind, however, that canyons and valleys channel wind, too, and it is entirely possible—probable in fact—that winds will roar up or down the canyon or valley floor no matter what time of day it is.

SNOWFIELDS AND GLACIERS

In mountain environments such as the Wind River Range in Wyoming, the Cascades of British Columbia, Washington, and Oregon, and the Sierra of California, the presence of year-round snowfields or glaciers creates localized wind conditions in the valleys directly below them. These winds are caused by the cool air that flows off the glacier or snowfield, displacing the surrounding warmer air that hugs the valley floor. The winds are fairly gentle and warm and dissipate quite quickly, disappearing altogether within ¼ mile of the glacier or snowfield.

Winds originating from glaciers or snowfields will occur as long as there is a sufficient temperature differential between the surface temperature of the snow and the surface temperature of the ground below the lip of the glacier or snowfield. The cold wind that emanates from ice caves underneath a glacier also has a significant effect on the growing season of plants. In the Cascade Range near the Canadian border, I have camped below several ice caves where the growing season of plants and flowers consistently begins several weeks after those of plants only several hundred feet away. This is because the colder air near the ice makes the plants think the growing season has not yet begun. This knowledge works well if you are trying to get a photograph of flowering alpine plants, as I was, and miss the first bloom due to miserable photography weather, as I did.

ALPINE ENVIRONMENTS

Wind is a constant in alpine environments, chapping lips, dehydrating the body, cooling the skin, and generally governing how and where plants will grow. Local upslope winds, caused by the sun heating the earth in the lower

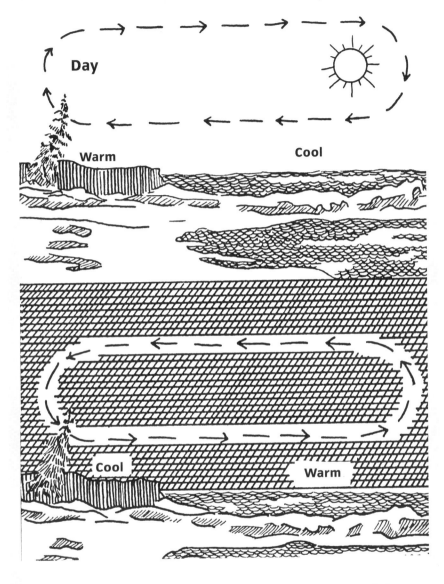

Figure 22. Land and sea air circulation differences between day and night.

elevations, begin around mid-morning. As the sun sets, the reverse effect is true as downslope winds begin, caused by the rapid cooling of alpine soils. Backpackers and mountaineers know not to camp in or near the bottom of subalpine valleys because of the intense cold that can be expected there—far colder than the tundra surface above.

Wind patterns in alpine environments can be altered by variations in the mountain terrain—rock outcrops, steep rock faces or slopes, and narrow passes or valleys. These variations in terrain help to create localized wind gusts, swirling eddies, and a phenomenon known as the Venturi effect. As wind gets forced through a narrow constriction, the localized air pressure drops and the wind accelerates. There is a pass in the Wind River Range where the prevailing wind on either side is relatively gentle. In the pass, however, the wind is felt almost gale force, blowing a tiny waterfall back up a rock face, evaporating it before it ever has a chance to hit the ground.

DESERTS

Thermally created low-pressure systems often develop in desert areas. As superheated air rises skyward, it creates a void, a low-pressure area that must be filled. The resulting effect is strong horizontal winds that rush in as the pressure attempts to equalize, creating dust devils and sandstorms.

LAKES AND OCEANS

Anyone who has spent anytime at all hiking or boating near the shore of a large body of water, such as the ocean or any one of the Great Lakes, is probably familiar with the daytime onshore winds and the nighttime offshore breezes. As with snowfields and glaciers, temperature differences between the land and sea must be sufficient for winds to occur. Unlike snowfields and glaciers, clouds will reduce the temperature contrasts. Winds typically are most common when the days are sunny and warm and the nights are clear.

Generally, onshore breezes will begin around mid-morning and continue until early to mid-afternoon. The wind will typically be steady and can result in a dramatic drop of onshore temperature, sometimes as much as 10 to 15 degrees F. Evenings and early mornings are usually calm. The offshore breeze often begins shortly after dusk and continues until early morning. See figure 22.

CITY ENVIRONMENTS

Cities create their own unique microclimates. Because of the high buildings, the congestion of people, and the concentrated production of pollutants, city

Figure 23. Thermal mountain over a city.

air and weather are often dramatically different from the suburban air and weather only a few miles away.

As the buildings absorb the sunlight, they trap the heat close to the ground. In addition, the particles of pollution that hang over the city serve to create a kind of seal, holding the heat in and not letting it escape into the atmosphere. This is why city temperatures are often 5 to 10 degrees F warmer than those in the surrounding countryside. This phenomenon has been named by some the thermal mountain (figure 23).

Cities also generate a great deal of moisture-laden air, which often leads to rainfall and fog within city boundaries but not in the surrounding countryside.

GENERAL GUIDELINES

Microclimates are as varied as the topography they occur in, but some generalities can be drawn.

Wooded terrain is generally warmer in winter and cooler in summer than open, treeless areas. Winds are lower and the humidity is higher in woods when compared to open areas.

Valleys have lower daily temperatures, wider temperature fluctuations, more frost and dew accumulation, and more fog frequency compared with conditions higher up in the hills or mountains. Wind velocity is also less in valleys when compared to that in higher elevations.

CHAPTER FOUR

Forecasting
Changes in Weather

USING AN ALTIMETER TO OBSERVE BAROMETRIC PRESSURE CHANGES

An altimeter/barometer (figure 24) registers relative changes in barometric (atmospheric) pressure. With the addition of a scale inscribed on the face, in meters or feet, the altimeter will take into account pressure changes that occur as an individual moves up or down in elevation. (An aneroid barometer, by comparison, merely measures air pressure.) Because of its compact nature, an altimeter is not ideally suited for reading the minute changes (recorded in 1/100 of an inch) used by the National Weather Service, but the standard ⅒-inch increments are adequate for amateur use.

The innards of an altimeter are simple, consisting of a vacuum chamber that contracts with high pressure (those pressures found closer to sea level) and expands with low pressure (pressures get lower the higher one climbs). Through the intricacies of gears and gizmos, the information is transferred

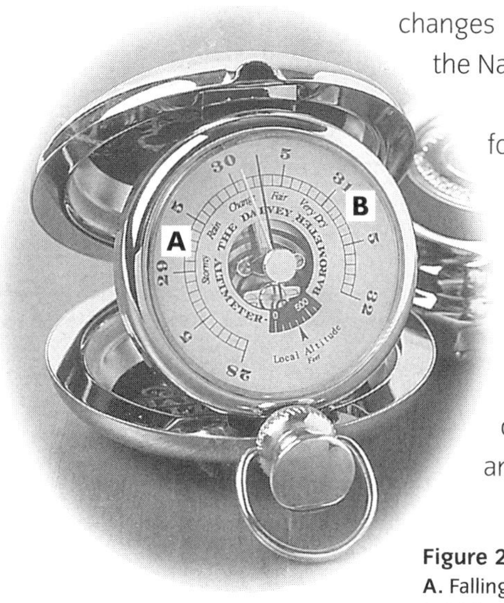

Figure 24. Altimeter/barometer.
A. Falling barometer
B. Rising barometer

and displayed by a needle on the face of the altimeter.

In general, when the weather service refers to a falling barometer (a lowering of the atmospheric pressure; see A in figure 24), the altimeter shows a rise in elevation. Conversely, a rising barometer (an increase in the atmospheric pressure; see B in figure 24) is read as a falling elevation on the altimeter.

How does this relate to changes in weather? Without getting specific, a rise in barometric pressure, demonstrated by a drop in your elevation reading while camped in one location, indicates improving or continued good weather. A drop in barometric pressure (a rise in elevation reading) is a precursor to deteriorating weather.

Be warned that any rapid change in barometric pressure, up or down, promises a change in the weather, and not always for the best. Slow changes usually indicate a stable weather pattern (dry or wet) that will last for a while.

PREDICTING WEATHER CHANGES USING WIND DIRECTION

The following are generalizations you can make about impending weather conditions by noting the wind direction (the direction the wind is blowing from, not to—a westerly wind is blowing from the west) and barometric changes. Reading barometric changes with electronic altimeters, such as the wrist-mounted Suunto Vector watch, is easier than with the old-style altimeters, as there are separate barometric functions and even hourly change-tracking functions. Both styles will afford the same information, though, if you remain observant.

Winds from the southwest to northwest

- If the barometric sea-level pressure is above 30.10 and holding steady, it indicates fair weather with little change for the next twenty-four hours.

- If the barometric sea-level pressure is around 30.10 but rising quickly, it should remain fair for about twenty-four hours, but rain and a warm front are on the way within forty-eight hours.

- If the barometric sea-level pressure is around 30.20 but falling slowly, then it should remain fair for the next twenty-four hours, but temperatures will likely rise.

Winds from the south to east

- If the barometric sea-level pressure is around 30.20 but falling rapidly, expect rain within twenty-four hours, maybe even as soon as within the next twelve hours.

- If the barometric sea-level pressure is around 29.80 but falling rapidly, a severe weather front is moving in, followed by an expected temperature drop.

Winds from the southeast to northeast

- If the barometric sea-level pressure is around 30.10, but it is falling slowly, expect rain within the next twelve hours.
- If the barometric sea-level pressure is around 30.10, but it is falling rapidly, expect rain and increasing wind over the next twelve hours.
- If the barometric sea-level pressure is around 30.00, but it is falling slowly, expect precipitation that will last up to forty-eight hours.
- If the barometric sea-level pressure is around 30.00 and falling rapidly, rain with accompanying high winds will prevail over the next twenty-four hours.

Winds from the east to northeast

- If the barometric sea-level pressure is around 30.10 but falling slowly, you can anticipate precipitation within twenty-four hours in the winter months and precipitation with light winds within forty-eight hours in the summer months.
- If the barometric sea-level pressure is around 30.10 but falling rapidly, precipitation with increasing winds will be the pattern over the next twenty-four hours.

Winds from the east to north

- If the barometric sea-level pressure is around 29.80 but falling rapidly, anticipate a severe weather front with accompanying heavy precipitation and possibly gale-force winds.

Winds blowing to the west

- If the barometric sea-level pressure is around 29.80 but rising rapidly, expect clearing and colder conditions.

ESTIMATING WIND SPEED

Devised by Sir Francis Beaufort in 1805, the wind-speed chart, now known as the Beaufort scale of wind force, is an excellent and extremely accurate method of estimating wind velocity.

THE BEAUFORT SCALE

Wind Speed (mph)

0–1	Smoke rises straight up; calm
1–3	Smoke drifts
4–7	Wind felt on face; leaves rustle
8–12	Leaves and twigs constantly rustle; wind extends small flags
13–18	Dust and small paper raised; small branches moved
19–24	Crested wavelets form on inland waters; small trees sway
25–31	Large branches move in trees
32–38	Large trees sway; must lean to walk
39–46	Twigs broken from trees; difficult to walk
47–54	Limbs break from trees; extremely difficult to walk
55–63	Tree limbs and branches break
64 on up	Widespread damage with trees uprooted

WINDCHILL INDEX

The windchill index chart demonstrates the actual cooling effect on bare skin exposed to air. Although understanding windchill has little to do with being able to predict weather patterns, it is important to realize the impact wind and temperature have on your body so you can make safe decisions regarding dress and travel.

HEAT INDEX

The heat index was devised by the National Weather Service. Using a rather complex formula and taking into account both air temperature and relative humidity, the heat index gives an indication of how hot it really feels to the skin—useful when planning your level of outdoor activity for the day.

To determine the heat index, look at the heat index chart on the next page. As an example, if the air temperature is 96 degrees F (found on the left side of the table) and the relative humidity is 55 percent (found at the top of the table), the heat index—or how hot it really feels—is 112 degrees F. This number is at the intersection of the 96 degree row and the 55 percent column.

HEAT INDEX °F

Temp.	Relative Humidity (%)												
	40	45	50	55	60	65	70	75	80	85	90	95	100
110	136												
108	130	137											
106	124	130	137										
104	119	124	131	137									
102	114	119	124	130	137								
100	109	114	118	124	129	136							
98	105	109	113	117	123	128	134						
96	101	104	108	112	116	121	126	132					
94	97	100	103	106	110	114	119	124	129	135			
92	94	96	99	101	105	108	112	116	121	126	131		
90	91	93	95	97	100	103	106	109	113	117	122	127	132
88	88	89	91	93	95	98	100	103	106	110	113	117	121
86	85	87	88	89	91	93	95	97	100	102	105	108	112
84	83	84	85	86	88	89	90	92	94	96	98	100	103
82	81	82	83	84	84	85	86	88	89	90	91	93	95
80	80	80	81	81	82	82	83	84	84	85	86	86	87

Keep in mind that the heat index values were devised for shady, light wind conditions. Standing in the sun will increase heat index values by up to 15 degrees F. Strong winds with very hot, dry air further increase the heat index and can dramatically increase the risk for heat-related illness.

PUTTING IT ALL TOGETHER

Combine the altimeter reading with wind direction, temperature, and cloud conditions to predict weather. Some generalized examples are as follows, although keep in mind that if predicting the weather were as simple as the following examples appear to make it, weather forecasters would never be wrong.

Learn to use the wind direction and your reading of the altimeter's fluctuations of elevation when stationary to aid in anticipating upcoming weather changes.

WINDCHILL INDEX

Temp	40	30	20	10	0	-10	-20	-30
5	37	27	16	6	-5	-15	-26	-36
10	28	16	2	-9	-22	-31	-45	-58
15	22	11	-6	-18	-33	-45	-60	-70
20	18	3	-9	-24	-40	-52	-68	-81
25	16	0	-15	-29	-45	-58	-75	-89
30	13	-2	-18	-33	-49	-63	-78	-94
35	11	-4	-20	-35	-52	-67	-83	-98
40	10	-4	-22	-36	-54	-69	-87	-101

(left axis label: Wind mph)

Warm Front

Incoming storm: The barometric pressure will fall steadily (the altimeter elevation rises). Wind is out of the southeast or northeast and increasing in speed. Cirrus will give way to altostratus then nimbostratus. As the clouds thicken, precipitation begins to fall, increasing in intensity as the front moves through. Temperature will increase gradually.

Outgoing storm: The barometric pressure will level off. Wind direction changes to come more from the south or northwest. Nimbostratus gives way to stratocumulus. Precipitation lightens to showers and drizzle.

Cold Front

Incoming storm: The barometric pressure will fall (altimeter elevation rises) slowly at first, more rapidly as the storm approaches. Wind will be from the south or northeast. Cumulus gives way to cumulonimbus. Brief but intense showers or hail. Little change in temperature.

Outgoing storm: The barometric pressure will rise sharply (altimeter elevation falls). Wind direction will shift to the north or northwest and become gusty. Cumulonimbus will begin breaking up, with partial clearing. Showers with intermittent thunderstorms followed by rapid clearing. Temperature will drop rapidly.

Occluded Front

Incoming storm: Barometric pressure falls steadily. Winds typically from the east or northeast, sometimes from the southeast with velocity increasing. Cirrus giving way to cirrostratus, then altostratus, then nimbostratus. Steady precipitation increasing as storm moves through. Temperature rises slowly.

Outgoing storm: Barometric pressure rises steadily. Wind from the southwest or north and decreasing. Stratocumulus to altocumulus with slow clearing. Precipitation tapering off. Temperature falls slowly.

USING WIND DIRECTION TO PREDICT WEATHER

The following information has been adapted from National Weather Service data regarding average conditions within the United States.

Winds from the Northwest

When winds are from the northwest and the weather has been clear, you may anticipate twenty-four hours of continued clear weather if your altimeter is steady or shows a slow drop in elevation (a rise in barometric pressure). If it has been raining and you read a drop in elevation, weather should clear in several hours. In all cases expect lower temperatures.

If the altimeter shows a rise in elevation (a drop in barometric pressure), expect it to remain clear for twenty-four hours if it has been clear, or expect the weather to change if it has been stormy.

Winds from the Southwest

When the winds are from the southwest and the weather has been clear, you may anticipate continued fair weather for twelve to twenty-four hours if your altimeter shows a drop in elevation (a rise in barometric pressure). If it has been stormy and you read a drop in elevation, the storm should pass in six hours.

If the altimeter shows a rise in elevation (a drop in barometric pressure), you may predict rain within the next twelve hours if the weather has been clear, or expect increasing rain with possible clearing after twelve hours if the weather has been stormy.

Winds from the Southeast

If the weather has been clear and the winds are from the southeast, you may plan on continued fair weather if the altimeter shows a drop in elevation (a rise in barometric pressure). If the weather has been stormy, anticipate clearing weather to follow soon if the altimeter shows a drop in elevation.

A rise in elevation shown on the altimeter during clear weather would indicate impending rain within twelve hours and possible high winds. If the weather is already stormy, a rise in elevation indicates an increase in severity of the storm followed by clearing within twenty-four hours.

WEATHER RULES OF THUMB

Clip out the following weather guidelines and take them with you. Combined with all the observation techniques and skills you are acquiring, they will give you a solid foundation from which to make accurate predictions concerning weather trends over any twenty-four-hour period. Add your own notes as you wish.

It is going to be FAIR when one or more of the following occur:

Wind is blowing from the west or northwest.

The barometric pressure remains steady or rises slowly.

Fair weather cumulus clouds dot the sky.

Early morning fog evaporates (gets burned off) by noon.

It is going to RAIN OR SNOW when one or more of the following occur:

The barometric pressure falls.

Cumulus clouds begin to develop vertical columns.

There is a halo around the moon.

Cirrus clouds begin to thicken and the cloud ceiling lowers.

The sky darkens.

A south wind increases in velocity.

A wind shifts in a counterclockwise direction, indicating a low. (A north wind shifting to west and then to south is a prime example.)

Expect the weather to clear when:

The barometric pressure begins to rise quickly.

South winds shift to the west.

Cloud ceiling begins to lift.

Anticipate cold temperatures when:

The night is clear, there is no cloud cover, and there is virtually no wind.

The barometric pressure rises in winter, or anytime in front of an oncoming system of clouds or an incoming cold front.

Winds from the Northeast

When the winds are from the northeast and the weather is clear, expect continued fair but cool weather if the altimeter shows a drop in elevation (a rise in barometric pressure). If the weather has been stormy, expect it to clear and the temperature to drop.

If the altimeter rises in elevation while the weather is clear, expect stormy weather within twelve to twenty-four hours. If the weather is stormy, expect very heavy rain coupled with severe gale-force winds and much colder temperatures.

CHAPTER FIVE

Nature's Signs

FORECASTING FROM LEGENDS AND LORE

Through the centuries humans have studied nature diligently to try and make sense of and anticipate weather changes. Long before the advent of sophisticated weather paraphernalia, humans read all the available signs. They listened to the animals, the plants, the wind, the earth. They used their eyes, their ears, their noses and their sense of taste. Not surprisingly, records show that these ancient ways of forecasting are almost as reliable as modern methods. Arm yourself with weather tools if you will, but don't ever close your eyes and ears to the signs of nature all around.

Morning or Evening Sky

"Red sky at night, sailor's delight. Red sky in morning, sailors take warning." This means that if the clouds take on a reddish hue in the morning, one can expect rain by the end of the day. If the evening sky is red, the weather will probably remain clear the following day.

Geese

Geese have a reputation for not flying before a storm. Biologists theorize that this is because they have a harder time getting airborne in low-pressure (thinner air) conditions. Or maybe they're just a lot smarter than we give them credit for.

Coffee

Grizzled outdoorspeople swear by the bubbles-in-the-coffee method of forecasting. Perhaps this explains why so many of them spend hours staring into a steaming mug of java. Remarkably, this method does seem to work and is attributed to the way pressure affects the meniscus, or surface tension, of the coffee. In high pressure the surface is rounded, like a globe. In low pressure the surface is concave, so naturally bubbles head to the highest point on the coffee's surface, the edges of the cup. For coffee forecasting to work, the brewed coffee must be strong. Instant coffee won't do the job, since there aren't enough oils to create satisfactory surface tension. Pour the coffee into a mug (vertical sides work best; venerable Sierra cups don't work as well). Give the coffee a good stir or two and watch the bubbles form. If they scatter this way and that and then form near the center, fair weather. If they cling to the sides of the cup, a low-pressure system is setting in, and rain is possible.

Mosquitoes and Blackflies

Anyone who has camped beside a body of freshwater prior to a storm's arrival knows that mosquitoes and blackflies will inevitably swarm and feast heavily up to twelve hours before the storm hits. You can bet the storm is almost upon you when the feasting frenzy subsides and the little buggers seemingly fly away to hide approximately one hour before the weather turns.

Fog

If the dawn is gray and there is fog in the valleys, the weather that day will be clear.

Frogs

Frogs of all shape and color have an uncanny tendency to increase their serenading several hours before a storm arrives. The reason they do this is not so much their reliability as weather forecasters but because the increased humidity in the air from an incoming storm allows them to stay comfortably out of the water for longer periods—their skin must be kept moist at all times.

Bees

A friend of mine, who is an amateur beekeeper, once told me that bees can sense inclement weather and will stick closer to the hive when the weather is about to take a turn for the worse.

Springs, Ponds, and Caves

Natural springs seem to flow out of the ground at a higher rate when a storm is approaching. This has been attributed as a natural response to a lowering of the barometric pressure.

The same lowering of barometric pressure will cause ponds, those with a lot of vegetative decay at their bottom, to become momentarily "polluted" as the decaying scum rides to the surface on marsh gases.

Stand near a cave's entrance during a storm's approach and you will become very aware of an outrushing of air. This is also due to a lowering of barometric pressure, became the air pours out of the cave to equalize the outside pressure. Just because a cave is breathing, however, doesn't mean a storm is on the way. Caves often breath naturally in response to outside temperatures.

Hair

Although not everyone is affected in quite the same manner, hair does react to changes in humidity. Because hair is a reliable indicator of good or bad weather, some instruments that measure humidity, called hygrometers, use hair as the primary working element. Like rope, hair tends to contract when it is damp and relax when it is dry. Straighter hair means dry weather. Wavier or curlier hair means wetter weather.

Halo around the Sun or Moon

In summer the sight of a hazy halo, or corona, around the sun or moon is a good indication that a change in the present weather pattern is in the forecast, most often for rain.

Frost and Dew

The presence of heavy frost or dew early in the morning or late in the evening is a fairly reliable indicator that up to twelve hours of continued good weather may be expected.

Canvas, Hemp, Wood–Ax Handle

Like human hair, hemp and canvas contract when it gets humid and stretch as it dries out. Woodsmen in the olden days noticed humidity changes as their wood-ax handles swelled during higher humidity (incoming rain) and loosened during drier weather.

Seagulls

Sailors know that if seagulls are clustering on the beaches and not flying over the ocean, it is a good indication that a fairly strong storm is about to blow in and boats had best stay in the harbor.

Songbirds

There is a theory that some birds tend to sing loudest right before a storm's arrival. Of course, this assumes that you have been carefully monitoring the volume of the birds in your area all day long. There is an equally popular theory that songbirds become silent right before a storm. I guess the real dilemma is which songbirds are going to sing what song and when. My suggestion is to take your pick, but keep a weather eye skyward just in case.

Colors of the Sky

Tints and hues of green, yellow, dark red, or a grayish blue indicate precipitation and, quite often, accompanying winds.

Cattle

Farmers have for years looked to their cattle to give them reliable indications of changing weather. It seems that cattle will herd together in lower elevations and off exposed hills when the weather is about to take a shift for the worse.

Wind

"Wind from the south brings rain in its mouth." If you look back to chapter 1 and the discussion about winds in the Northern Hemisphere, you will note that low-pressure systems create cyclonic winds that rotate in a counterclockwise direction. Since low-pressure systems are frequently associated with rainstorms, the rhyme proves quite accurate. Counterclockwise wind rotations create wind that blows from the south, wind that brings in the rain.

High-pressure systems are often associated with clear or clearing weather, with clockwise rotating winds. Keep aware of wind directions and you will keep your finger on the weather's pulse—is it beating fair or foul?

Another theory that holds a certain amount of credibility states that if the wind comes first from the north, then the west, and finally the south, rain is imminent. Winds that begin in the south, shift to west, and then back to north indicate the storm or cloud cover is clearing.

Smell

In the Midwest and the Great Plains, the smell of rain has a definite meaning. As a storm approaches, with an accompanying drop in pressure and rise in humidity level, the ground begins to give off a rich and sweet odor that almost smells like freshly mown hay.

Sound

Sound does, apparently, travel farther when a storm approaches. Higher humidity helps to carry sound waves, as does the wind from an approaching low.

Plants

Dandelions are often credited with being cold-front predictors, as they will close up their flowers when the temperature drops below 50 degrees F.

Observe a field of clover and you will note that the clover rolls its leaves up at the approach of strong winds. Since strong winds often precede a storm, clover is known as a reliable indicator of incoming weather. Still, one could argue that it is the wind and not the clover that is the harbinger of weather tidings.

Campfire Smoke

By observing the smoke from your campfire (figure 25), you can monitor what pressure system is in the area, low or high. If the smoke from the fire hangs low to the ground and seems to dissipate into surrounding tree branches, that means a low-pressure system is here and rain is possible. If the smoke rises in a straight, vertical column, high pressure rules and fair weather may be anticipated.

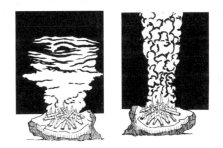

Figure 25. Campfire smoke patterns.

The Calm before a Storm

If the wind has been blowing and clouds have been moving in rapidly for the past few hours and then, suddenly, the wind dies, head for cover because you are in for one heck of a gully washer.

Crickets

Believe it or not, you can actually estimate the air temperature in Fahrenheit by counting the number of chirps a cricket emits over a fourteen-second period and then add 40. If a cricket chirps 25 times over fourteen seconds, you add 25 to 40 and arrive at 65 degrees F. Scientific studies have proven that the crickets are correct within a degree or two more than 75 percent of the time. Keep in mind, though, that air temperatures closer to the ground where the cricket resides will differ by as much as several degrees from those found a few feet higher.

Deer, Big Horn Sheep, Elk

Wild grazing animals such as these, which are common in mountainous areas, will generally move down a mountain and into sheltered valleys as a storm approaches and will begin to move back up a mountain as a storm gets pushed out. The trouble with watching for this movement is that by the time you see it, chances are the storm is already bearing down and highly visible.

CHAPTER SIX

Backyard Meteorology

SETTING UP A WEATHER STATION

Setting up a backyard weather station is not only fun, it can be useful as well. Combine a little experience and the right equipment, and anyone can learn to safely predict weather changes at least twelve to twenty-four hours in advance. That can make the difference between knowing whether to pack a raincoat or sunscreen on your next jaunt around town.

With the increasing use of computerization, home weather stations have become more compact and high tech than ever before. Electronic weather stations (figure 26; see also appendix 3) allow the amateur observer to document all pertinent weather information from the comforts of his or her home, without even stepping a foot outdoors. A miniature computer registers

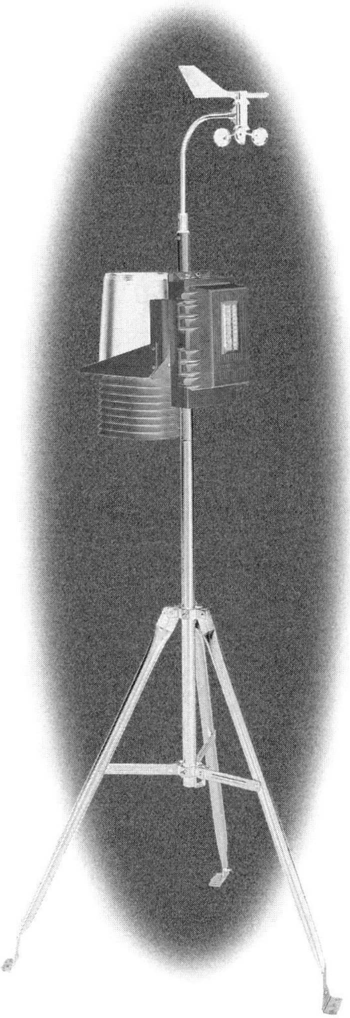

Figure 26. Sensor array for an electronic weather station.

Figure 27. A barometer.

Figure 28. Wind vane.

Figure 29. Rain gauge.

information regarding wind speed and direction, temperature, barometric pressure changes, and rainfall obtained from outside instrumentation mounted on a roof or other open area. Talk about your cozy weather forecasting. On the other hand, one could argue that computerization could lead to a decidedly detached approach to forecasting—nothing like being in the thick of things for true hands-on and experiential weather observation.

For a more traditional approach, create your weather station out in the open, with all the instruments close enough together to facilitate easy reading and recording of information. To get the most out of such a weather station (figure 30), you will want to assemble the following instruments: barometer (figure 27), wind vane (figure 28) and anemometer, and a thermometer. A rain gauge (figure 29) and hygrometer are accessories that are not absolutely necessary for basic weather forecasting. They are most useful, however, if you wish to assemble a complete weather station and have the space.

UNDERSTANDING A WEATHER MAP

The weather maps that you see every day in the newspapers are usually created from information provided by the National Weather service maps. While the newspaper weather maps are not the most reliable indicators of long-range weather forecasts, with practice it is possible for anyone to become somewhat adept at anticipating what the weather holds for twenty-four-hour periods. *USA Today* and the *New York Times* offer the best overall reproductions of a weather map.

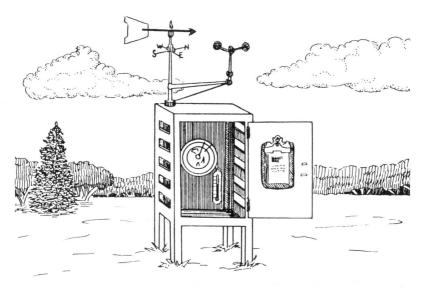

Figure 30. A weather station. A box with well-ventilated (slotted) sides, set on stilts away from buildings, stores weather gear (barometer and notebook) inside. A wind vane and an anemometer are mounted on top.

In order to read a weather map (figure 31), it is important to understand the symbols. The thin, black, wavy lines that cross the map, resembling contour lines common on topographic maps, are isobars. Isobars are lines of equal barometric pressure. These lines illustrate pressure patterns that control air flow. Air moves fastest where the lines are closest.

Four other lines on a map have geometric symbols—one all triangles, one arches, one an alternating combination of triangle and arches facing the same direction, and the fourth an alternating combination of both triangles and arches facing opposite directions. A line with all triangles indicates a cold front. A line with all arches indicates a warm front. The alternating combination of triangle and arches facing the same direction indicates an occluded front. The alternating combination in opposite directions shows a stationary front, a front that is not moving. The direction in which either a cold or warm front is moving is indicated by the direction the triangles or arches are pointing.

Wind directions and speeds are designated by tiny symbols that look remarkably like musical notes—a little circle with a tiny flag attached. The arm, or flag, of the symbol points in the direction the wind is coming from.

Shaded areas of the map indicate areas where precipitation is falling. Check the map's key to determine how rain or snow is differentiated.

DEVELOPING A ROUTINE

Every day, at the same time each day, establish a weather observation routine that never varies. You don't need to spend more than ten to fifteen minutes each time, but consistency is the key to accuracy. Be sure to document your observations and weather comments in a logbook. That way, you can go back and establish a weather history from which you can learn the patterns that lead to various weather systems.

1. Look to the sky. Observe and identify as many cloud formations as you can. Note the direction they are moving and whether or not they appear to be part of a system. Are they forming or breaking up? What is their sequence of formation?

2. Using your wind vane and anemometer (or rely on the Beaufort scale for estimating), determine the wind speed and direction. Have they changed from yesterday? How? Is the wind increasing or decreasing? Does the wind direction, coupled with the formation sequence of clouds, indicate a front moving in?

3. Check the temperature and the barometric pressure. Is the temperature going up or down? Why? Is the barometric pressure going up or down, or is

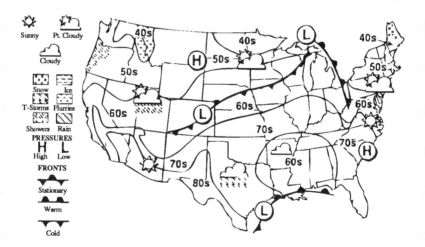

Figure 31. Weather map.

it holding steady? Why? How does this information mix with what the cloud formations tell you?

4. Check the rain gauge for precipitation levels.
5. Enter all the information in your log. Your log can be set up any way you want, but it should include separate columns for the following: date, time, wind direction and velocity, temperature, barometric pressure, precipitation, forecast, and changes. Changes are for you to fill in after your forecast if what you predicted doesn't pan out. Don't forget to note why what you predicted didn't happen and what happened to the weather to change the conditions. It is also a good idea to have a special column where you can note conditions such as frost, dew, or fog.

Appendix 1: Glossary

acid rain: Rain that contains acid pollutants produced by factories and automobiles. Acid rain leaves lakes infertile, unable to support fish or plant life. It also kills vegetation and trees.

adiabatic cooling: Cooling of the air due to a rise in elevation.

adiabatic warming: Warming of the air due to a lowering in elevation.

anemometer: An instrument for measuring the velocity or miles per hour that the wind is traveling. Note: most weather bureaus report the wind speed in nautical miles per hour, or knots, as they are a universally understood unit of measurement.

atmospheric pressure: The weight of a column of air on the earth's surface measured by a barometer in inches or centimeters.

barometer: An instrument used to measure atmospheric pressure.

chinook winds: Relatively strong winds, usually warm and dry, that occur in valleys and canyons on the leeward side of mountain ranges.

coalescence: The creation and growth of a droplet of rain large enough to fall to earth through the atmosphere.

Coriolis effect: The effect on air movements caused by the earth's rotation. Air movement curves to the right in the Northern Hemisphere and to the left in the Southern Hemisphere.

corona: A colored circle seen when the light from the sun or moon passes through a thin veil of high moisture-laden clouds. Most often, the corona looks like a yellowish/white disk surrounding the moon or sun, sometimes with an inner blue border and a dirtier brown outer edge.

dew point: The temperature at which the relative humidity equals 100 percent. This causes moisture in the atmosphere to condense into rain.

downdraft: Flow of air downward due to a lowering in air temperature, most often caused by falling precipitation.

drizzle: A very light rainfall. Droplets are uniform in size, are quite small, and typically fall from stratus clouds.

front: The boundary between two air masses, named for the dominant air mass moving in—cold, warm, or occluded front.

graupel: A mass of frozen droplets of precipitation that is lumpy and soft. Frequently falls in conjunction with a severe thunderstorm. Softer than hail, more bulky than snowflakes.

green flash: A seldom-viewed and somewhat mystical occurrence at sunset that is seen immediately following the disappearance of the red rim of the sun beneath a clearly defined horizon. The green flash is created by sunlight refracted through various densities of atmospheric layers and lasts only a few seconds.

hail: Moisture particles frozen in areas of intense updrafts, often falling earthward only to be tossed up once again by a strong updraft. Formed most typically in large, billowing cumulonimbus clouds. Large hail is formed as hail particles begin to thaw and then refreeze after gathering more moisture. Hailstones as large as baseballs have been recorded.

halo: A halo is caused by the presence of ice crystals in cirrostratus clouds that fall between the light source and the observer. It may sometimes be viewed surrounding the sun or moon.

high pressure: An air mass with greater-than-normal air pressure, often referred to as an anticyclone.

hygrometer: An instrument used to measure the amount of moisture in the air, otherwise known as humidity.

isobars: Lines on a weather map that indicate lines of equal barometric pressure.

jet stream: The name given a band of rapidly moving air caused by cold polar air moving south and warm tropical air moving north. The jet stream heavily influences storms and is frequently referred to by weather forecasters. The jet stream is most powerful in winter, since global temperature contrasts are at their greatest.

low pressure: An air mass with lower-than-normal air pressure, often referred to as a cyclone.

mist: Extremely small raindrops that fall at a steady rate from stratus clouds.

orographic lift: When physical barriers such as mountain ranges force air currents to rise, this is known as orographic lift.

ozone: A beneficial, naturally occurring component of the upper atmosphere. A shell called the ozone layer shields the earth from ultraviolet rays.

rain: Larger than drizzle yet smaller than showers, rain is made up of consistently sized droplets of moisture that fall steadily from nimbostratus clouds.

rainbow: When the sun's rays pass through a drop of water, the light that passes through the extreme upper and lower edges of the drop is bent by refraction, creating the colors the eye sees as a rainbow.

raindrop: Contrary to popular myth, raindrops are not shaped like a teardrop. Instead, they are round in shape when small and somewhat flatter in shape when large. The shape is determined by friction with air.

relative humidity: The amount of measurable moisture in the air.

rime: Originating from a supercooled cloud passing rapidly over a mountain summit, moisture droplets freeze upon impact, creating wonderfully beautiful ice formations that take on a featherlike quality. Ice formations up to several feet long protruding from tree branches and rock faces are not unusual. Usually only viewed on exposed rock faces and other objects at or near mountain summits.

saturation point: Also known as 100 percent humidity, saturation point is the point at which the air cannot hold any more moisture.

showers: A fall of rain made up of the largest of the droplets of moisture falling from clouds. Shower droplets are large in size and fall in dense sheets from heap clouds such as cumulonimbus.

sleet: Precipitation that is partially frozen and partially liquid. Formed predominantly when drizzle falls through cold air and begins to freeze from the outside first.

snow: Frozen precipitation that may take many crystalline forms, depending greatly on the temperature and moisture content of the clouds in which the snow formed.

sun dog: Unusual bright spots, or second suns, that can appear on both sides of the sun through refraction involving the viewer, the sun, and ice crystals in the air.

thermal mountain: The invisible dome of hot air that builds over a city, prevented from escaping by the particles of dirt and pollution that hang over the city.

thermometer: An instrument used to measure fluctuations in air temperature.

updraft: An upward flow of air from a warm surface, usually caused by the earth's surface being warmed by the sun.

Venturi effect: Wind is forced through a constriction (a narrow mountain pass, narrow streets between skyscrapers), causing a drop in air pressure and an associated acceleration of wind.

American Weather Observer, a monthly publication of the Association of American Weather Observers; www.aawo.net.

It's Raining Frogs and Fishes by Jerry Dennis, with drawings by Glenn Wolff (Harper Paperbacks, 1993). The only book on weather and natural phenomena that you'll actually enjoy reading before going to bed. Packed with loads of information relating to the origin of rain, how storms begin, etc.

National Audubon Society Field Guide to North American Weather; by David Ludlum (A.A. Knopf, 1991). One of the most complete guides to understanding the weather I have seen. It includes more than 378 color photographs illustrating cloud types, storms, precipitation, and more.

National Weather Service Publications
Various useful publications may be ordered from the National Weather Service by writing the Superintendent of Documents, U.S. Government Printing Office, Washington, DC 20402. Ask initially for a listing of publications and the associated price list. Some of the more useful pamphlets are: *Climates of the United States, Cloud Code Chart, Instruments Used in Weather Forecasting,* and *Weather Forecasting.* A particularly outstanding book on weather, also available from the U.S. Government Printing Office, is *Aviation Weather for Pilots and Flight Operations Personnel,* second ed., coauthored by the FAA and the Department of Commerce.

1001 Questions Answered about the Weather, by Frank Forrester (Dover Publications, 1981). The title says it all.

Peterson First Guide to Clouds and Weather, by John A. Day and Vincent Schaefer (Houghton Mifflin Company, 1998). A pocket guide to understanding weather.

The Weather Companion, by Gary Lockhart (John Wiley & Sons, Inc., 1988). A marvelous and detailed look at meteorological history, science, natural legends, and folklore.

Weather Station, by Deni Brown (DK Publishing Inc., 1998). A fundamental guide.

Weathering the Wilderness, by William F. Reifsnyder (Sierra Club Books, 1982). An excellent guide to specific weather patterns and detailed weather information for regional wilderness climates.

Weatherwise. An excellent magazine focused on meteorology and climatology with articles on technology, history, culture, art, and society as each relates to weather; www.weatherwise.org

http://lwf.ncdc.noaa.gov/oa/ncdc.html
The National Climatic Data Center offers a wealth of historical weather and environmental data as well as records.

http://pollen.com/Pollen.com.asp
This resource offers information on the most prevalent pollens, broken down by zip code—very useful if you suffer from allergies.

www.accuweather.com
AccuWeather is comprehensive, the only Web site I know of that is consistently mentioned as often if not more than the venerable Weather.com. I've been amazed at the accuracy of its local forecasting—down to the hour when a weather event will happen.

www.adventurenetwork.com
If you have a question you simply cannot find the answer to in this book or in any of the other resources, or you simply want to contact me, the author, feel free to visit my Web site and e-mail me from there. I'll do my best to answer any question you have relating to the outdoors and wilderness weather as quickly as possible.

www.ncep.noaa.gov
The National Centers for Environmental Protection offers a global perspective of the weather and provides a broad look at meteorology and environmental forecasting.

www.noaa.gov/wx.html
The National Weather Service provides weather, hydrologic, and climate forecasts and warnings for the United States, its territories, adjacent waters, and ocean areas for the protection of life and property and the enhancement of the national economy.

www.weatheraffects.com
Weather Affects is a one-stop resource of links and information related to weather forecasting.

www.weather.com
It's the Weather Channel of cyberspace, and it's chock-full of forecasts for local regions as well as tons of educational material that'll make any amateur meteorologist begin salivating.

Appendix 3: Suppliers of Weather Instruments

American Weather Enterprises: www.americanweather.com

Carolina Biological Supply Co.: www.carolina.com

Cloud Chart, Inc.: www.cloudchart.com

Davis Instruments: www.davisnet.com

Delta Education: www.delta-ed.com

Edmund Scientific: www.edsci.com

eNasco Science: www.enasco.com/top/523/

Fisher Scientific Co.: www.fishersci.com

Forestry Suppliers, Inc.: www.forest-suppliers.com

Frey Scientific Co.: www.freyscientific.com

Hubbard Scientific: www.hubbardscientific.com

Kestrel Pocket Weather Meters: www.nkhome.com/ww/wwindex.html

Northeast Discount Weather Catalog/Viking Instruments: www
  .discountweather.com

NovaLynx Corp.: www.novalynx.com

Oregon Scientific: www.oregonscientific.com

Radio Shack: www.radioshack.com

Robert E. White Instruments, Inc.: www.robertwhite.com

Schoolmasters Science: www.schoolmasters.com/weather.html

Sargent-Welch VWR Scientific Products: www.sargentwelch.com

Wind and Weather: www.windandweather.com

Index

A

adiabatic fluctuations (cooling or warming), 13, 24, 25, 26
alpine environments, 30–32
altimeter, 34, 35, 38, 39
altocumulus, 15, 20
altostratus, 4, 5, 19, 20
anemometer, 50
aneroid barometer, 34
atmospheric pressure, 9, 34

B

barometer, 50
barometric pressure, 2, 3, 15, 35, 36, 39, 40, 41, 42
Beaufort scale of wind force, 36, 37
bees, 44
blackflies, 44

C

campfire smoke, 47
canvas, 45
cattle, 46
caves, 45
chinook winds, 26–27
cirrocumulus, 15, 16, 18, 21
cirrostratus, 4, 18, 20, 21, 22
cirrus clouds, 3, 4, 15, 16, 21, 22
city environments, 32–33
clouds, 14–23
clover, 47
coalescence, 10
coffee, 44

cold front, 4–5, 14, 20, 39
condensation, 13–14, 24–26
continental polar, 1, 2
crickets, 48
cumulonimbus, 5, 6, 16, 17, 21
cumulostratus, 5, 20
cumulus, 6, 15, 18
cumulus congestus, 18, 21

D

dandelions, 47
deer, 48
depressions, 3
deserts, 32
dew, 10, 33, 45
dew point, 4
downdrafts, 7

E

elk, 48
evaporation, 13, 14

F

fair-weather cumulus, 17, 18, 21
fog, 18, 33, 44
frogs, 44
fronts, 2. *See also* cold front, occluded front
frost, 10, 33, 45

G

geese, 43
glaciers, 30

H

hail, 23
hair, 45
halo, 15, 45
heat index, 37–38
hemp, 45
high pressure, 2, 7–8, 26, 41, 44, 47
hurricanes, 8–9
hygrometer, 50

I

incoming storms, 39, 40

L

lakes, 32
lenticular clouds, 19, 20
lightning, 6, 7–8, 18
low pressure, 2, 8, 26, 43, 46

M

mare's tails, 15
maritime polar, 1
maritime tropical, 1
microclimates, 24, 33
mosquitoes, 44
mountains, 27–30, 32

N

National Weather Service, 9, 34, 40, 50
nimbostratus, 4, 5, 19, 20, 21

O

occluded front, 3, 5–6, 40
oceans, 9, 32
outgoing storm, 39, 40

P

plants, 47

ponds, 45
precipitation, 1, 3, 4, 5, 6, 10, 13, 14, 20, 26
pressure systems, 2

R

radiation, 10, 27
rain, 10
rain gauge, 50
ridges, 3

S

Santa Anna winds, 26
scud clouds, 15
seagulls, 46
seasons, 12
sheep, 48
sky, 43, 46
smell, 47
snow, 10
snowfields, 30
songbirds, 46
sound, 47
springs, 45
stratus, 18–20
swelling cumulus, 18, 21

T

thermal layer, 27
thermal mountain, 33
thermometer, 50
thunder, 6–8, 15
thundershowers, 5, 6, 7, 15
tornadoes, 9

U

updrafts, 6, 9

V

valley, 27, 29, 30
Venturi effect, 32

W

warm front, 9, 20, 39
water cycle, 13–14
weather map, 50–52

weather station, 49, 50
winds, 40, 42, 46
wind chill index, 37, 39
wind directions, 35–36, 40–42
wind speed, 36–37
wind vane, 50, 51
wood–ax handle changes, 45

About the Author

Recognized nationally for his poignant writing style, humor, and knowledge of outdoor equipment, award-winning journalist and author Michael Hodgson is constantly "seeking out the wilder side of a mountain" in search of a good story or a tough place to test gear. He is the copublisher of SNEWS, a subscription-only newsletter serving the outdoor and fitness industries, and a founding partner of www.GearTrends.com, an information site for new products and trends in the outdoor, snow sport, fitness, paddling, and bike markets. His www.adventurenetwork.com was recognized in 1999 as a *USA Today* hot site, a Featured Expert award winner, and a Golden Web award winner.

Formerly the gear editor for *Men's Health* and an editor for *Outdoor Retailer, Daily Exposure, Planet Outdoors,* and *Fogdog Sports,* he writes articles that have appeared in *Outside, Backpacker, Hooked on the Outdoors, Men's Journal, Adventure West, Field and Stream, Outdoor Life, Christian Science Monitor,* and other periodicals. He has written eighteen books on the outdoors, including *Compass & Map Navigator, Facing the Extreme,* and *Camping for Dummies.*